Get More Building For Your Money

ADVICE FOR A FUTURE BUILDING OWNER

By Lawrence "Larry" Riley

Construction Cost Cutting Consultants

Published by ConCoCuCo Publishing

2004

Re-edited 2009

Cover Art by Beldarr

www.beldarr.com

Author's Note – A Synopsis

To show my appreciation for buying my book, I am now going to save you the trouble of reading it. All you have to do is study the cover.

There are two buildings shown. Both are Medical Care Facilities housing about the same number of patients. Both were built in the same county at approximately the same time.

The top one was built using public money and was designed by an Architect, then built by the lowest bidder. The bottom one was built with private money, and was designed and built by the Owner.

*The top one cost easily **twice** as much per patient as the bottom one! Probably more than twice, because of the inefficient use of space dictated by several of the design factors. Plus – the continuing costs of utilities and maintenance are also going to be more!*

I will not bother to discuss the motivations involved in the designs. You probably already know some of them. If not, you may want to read the book anyway. I think you will find it not only revealing, but – at least mildly – entertaining.

Now look at a few of the details in the top picture: The obtuse and acute angled corners and cantilevered overhangs alone are enough to make a bean-counter cry (if he has read this book). But there is more – much more. The wall materials and the window patterns. And there are two sections of roof that rise in hexagonal pyramids to glass monitors that are capped with smaller hexagonal pyramids. Enough to make a steel detailer cry – while his boss rejoices.

I would comment on the high walls and the two stories, but the sprawling irregular floor plan that is built into a hillside is overwhelming enough. And, to me, the worst part is that this building is not "homey" – which would be my first criteria for a "care home".

So you can understand why I was thinking of labeling the top picture "What Were They Thinking??" Or maybe "Designed to Impress" while the bottom one would be "Designed to Be Used".

To complete this synopsis of the book, just in case you have decided not to read it, let's consider whom "they" were trying to impress and who "they" are. The first "they" are the county commissioners who are a committee charged with spending the county's money wisely. Or, so we should think. (Notice the word "committee".) In this case, they have to get a new medical care facility built. They choose an Architect by using some mysterious criteria. Now that Architect becomes the main player in "they". And whom does it most behoove him or her to impress? Their future clients! The project that they have somehow landed is destined to become part of their resume to land another project. An edifice to promote themselves, in my slanted thinking. All they have to do is get the board to approve the money.

And contrary to all economic logic, the more the building costs, the more the Architect makes.

*So, to complete this little synopsis, I will quote the last line of the book: **Design is where it's at!***

Introduction

This book is written to the prospective Owner of a new building. It would be most useful if read <u>before</u> the first steps are taken on the project. Unfortunately, as in most of life, we do not know we need help until we're in trouble. But I hope it can be of some use at any stage.

I was an estimator, designer and project manager for a very successful general contractor for over 35 years. My special interest was efficient design. I DETEST waste in any form! I can help you avoid a lot of waste and gain a lot of value in your next building project. Now that I am retired from the business and have nothing to fear, I can tell you some truths. You can believe me because I also have nothing to gain.

Preface

(This will be a repeat of the synopsis, since I added that later. I have decided to leave it in because it bears repeating.)

The cover says it all. These two buildings serve the same purpose, they are both nursing homes, though that may not be the politically correct term. The building built with public money cost at least twice as much per square foot as the private one. Also, I will guarantee that it has more square feet per bed. Wasted, unusable, undesirable square feet. It was obviously designed to impress, but just who, I'm not sure. It certainly is in the Architect's portfolio. And the politicians who were in office at the time probably are quite proud of their edifice. But it wasn't their money. In fact, the Architect made more by spending more. And do the patients, who are mostly confined to the interior, care what the exterior of the building looks like?

The building built with private money was designed to pay for itself. To be successful, the facility has to be inviting to patients and their families. This is the newest building built by this very successful Owner. Ironically, the less expensive building is surely more "homey", hence <u>better</u> suited to its purpose. For FAR less money!

Here's the Secret Already

I'm not going to make you wait until the end of the book to give you the secret. Of course, nothing is **really** a secret, and this is not either. You just need to realize the logic and apply the lessons you can learn from me.

Between 80 and 90% of the cost of a building are "hard costs". We hope. If you pay more than 20% in "soft" costs such as consulting and management fees, you're definitely going to be in trouble. "Hard Costs" in construction refer to the "bricks and mortar" – the materials that you see in the final structure and the labor it took to put them in place.

"Hard costs" are affected by two things: Size and DESIGN. The size is usually something that you, as Owner, decide. (There are ways to get the same performance from a smaller building, but the value of the facility will usually be defined by the size.) The design, however, is where the difference can be made.

And, therein lies the secret: **"DESIGN is where it's at!"** Design includes the materials used and the way they are used. Design includes the shape of the building and the appearance. Design includes the siting and the site work. **Every design decision affects the cost.**

The **design_er_** determines the performance and the quality – and the **COST.** The builder is obligated to build according to the designer's plan.

I could never figure out how an Architect could get away with blaming the Contractors for the high bids for **his** design. But they do! I even know of one instance that the bids for a small library were way over budget (the Architect's budget) and the Architect was quoted in the local paper as saying that "the contractors (plural – all of them) did not understand the drawings". That's the most self-damning statement I ever heard. [Incidentally, when bidding one of his jobs, we always added a "(his name) factor" because we knew how unreasonable he was to work for.]

It is the Architect's responsibility to put out drawings that can be understood. **Please don't EVER blame the builder for the high cost of a design.** Especially when you have competitive bids.

A Primer

There are three basic ways to get a building built. The first one I will discuss is the most conventional and has been used the longest. It is the method most people think of when they embark on a building project. And if you are planning to build a **Museum**, it really is the *only* way. (See below.) I will call it the SYSTEM.

But the system is flawed! **IT IS <u>BADLY</u> FLAWED!** It has been flawed for many decades and it appears that it is not going to get any better!

You need to know how the system works to protect yourself. First, let's take a look at what you are going to face.

To get a building built you need several things: money, land, a design, permits, a builder – and more money. You may think that you can be your own builder. But you probably can't borrow the money without prices from an established builder. You can't get prices from a builder without a design. The builder can't work without permits and the authorities don't issue permits without the seal of a "registered architect or engineer." These are just the high points of a complicated process. Besides, you probably should be concentrating on what you do best, your business – the business that needs a new facility. You'll spend enough of your time telling the designer what you want the facility to do.

Now let's look at how this age-old system works. I'm going to start with the traditional method used by private businesses and governmental institutions.

Simply put, the Owner hires an Architect who designs the facility, "draws up blueprints and specifications", "puts it out for bids", advises who the low bidder is, writes the contract, then supervises the construction and payments and does the final inspection – all for a percentage of the cost.

That's a percentage of the **Owner's** Money. There's something wrong with this picture. The *more* the Architect spends of your money, the more **he** makes. In fact, if he makes a mistake that necessitates a "change order", you not only pay the contractor for the work, you also pay the Architect to make the change – to correct his own mistake.

Before I start hearing from the Architects, I will admit I have known a exceptions to these conventions. I have even dealt with one architect that did not charge for his mistakes – he even paid the contractor to fix them! But that is indeed rare. So yes, you can find Architects that will agree to negotiate around some of these problems.

The Architect has no money at stake except for what he has spent in his office. If his design puts the project in jeopardy, he has little to lose. You can't fire him because no "brother" Architect will touch the job.

One of the popular misconceptions is that you should pick the Architect with the lowest percentage fee. Or a fixed fee. ***Hear me now or do not bother to read any further!*** The amount of the "fee" PALES against the cost of the design! And the cost of the contractors' fees also PALES against the cost of the design! **EVERYTHING** <u>PALES</u> AGAINST THE COST OF THE DESIGN!

The irony is that it is a lot more difficult to do a design that is cost efficient. And it is a lot more difficult to write specifications that encourage competition and still get the Owner quality. And time really is money. <u>Especially</u> in an Architect's office. So, ideally you would want a contract that rewards the Architect based on how much he can **save** rather than how much he can spend. And that is <u>extremely</u> difficult, especially since it will appear to be in conflict with your desire for a high quality facility.

Are you now thinking that you face an impossible task? Well, maybe not. Let me help by sharing some of what I learned in my career. First though, let me vent a little more. You will learn from that, too.

Have you noticed that I always capitalize the word Architect? Architecture has always been one of the "Professions". Architects have to have more schooling and job experience than Engineers and their testing and licensing requirements are quite strict. Their field is a broad combination of Art and Science. Their buildings are expected to be functional while the design is expected to be creative and unique. They are "Consultants". We pay them for advice. We trust them. We don't have much choice.

But we usually trust them too much. How many people do you know that would be smart enough and talented enough to do all the things we ask of our Architect? Smart enough to tell you what your building should look like and how it should perform AND then try to tell dozens of experienced craftsmen how to do *their* jobs and exactly what materials to use. And then let you expect the whole thing to work for more than the year that the contractor guarantees.

I don't know just where to put this in, so I guess now is as good a time as any. If you are building a MUSEUM where a unique design is foremost and money is no object, then by all means, you need an Architect. But that's **ARCHITECT**, all caps, as in I. M. Pei or Frank Gehry. Then, do you know what he (or she) will do? He will hire a structural consultant (he'll keep looking until he finds one that can make his outlandish design structurally feasible) and he will hire mechanical consultants and electrical consultants and site consultants and interior consultants and maybe even a roofing consultant (but probably not, even though that is what they usually need most – did you know that many of Frank Lloyd Wright's roofs leak?). Then they will probably "put it out for bids" and "let it to the lowest bidder". Do you see the irony in that?

Now we're on the right track. A *good* Architect will get help in the areas that he is weak. He will hire consultants that are specialists in their particular fields. A *real good* Architect will choose the best specialist for your facility. But an *exceptionally good* ARCHITECT will also have a working knowledge of the structural and mechanical disciplines and they will already be considered in the original concept of his design.

[Incidentally, that is when most of your money is spent (or can be saved) – when the original concept is developed. If you are already in trouble, it might be time to throw everything out and start over.]

There is still the concern that he may be building a monument to himself. That is almost irresistible for any designer because he has pride in his own creativity. (I was even guilty of that!) You may not mind since you will be the proud Owner. Except it is *still* your money. But the exceptionally good ARCHITECT can make a bold design statement without lavish costs, if he tries.

Now we're back to an incentive for him to save construction costs. But, as I said, that is **very** complicated.

So how <u>do</u> you select an Architect? I'm not sure that even I know although I was doing it for my own buildings and making recommendations to my customers. It is by far the most important and yet the most difficult part of the whole process.

I can tell you some stories and make some suggestions.

Once upon a time I had a customer who owned a chain of theaters. Eventually I did 26 projects for them in three states. We started with remodeling and additions and then finally we developed a prototype design for new buildings. The design was unique and quite "showy" but still very efficient and flexible. Every need and wish that the Owner had and every efficiency that we had learned were embodied in the design. Two were built and the Owner was very happy with them.

I had gone through three company presidents and the fourth was a young woman who inherited the business from her father. She wanted to make her own mark and that included rejecting our prototype design. (In her defense, I should not have been trying to prevent her from wasting her money on frivolous design. That was her prerogative.) Anyway, she commissioned an Architect to do a *new* prototype. He had designed her new corporate headquarters but had never done a theater.

After he had completely drawn the new design, I was asked to take look at it. I, reluctantly, but graciously (I hoped), and, of course, with great suspicion, went to the Architect's office, purportedly as the Owner's consultant.

When I asked where the seat layout was, he said the seating company would do that. So I asked how he had arrived at the auditorium dimensions (which should be determined by the seat layout) and he replied that the Owner had given them to him. Then I recognized some of the dimensions from our prototype, to the inch. Some, but not all. So the entire concept was flawed. The Architect knew nothing about theaters.

We built one of the prototype design facilities and I saved the Owner **$50,000** just by changing dimensions within the same footprint. And I added 30 seats which is significant since the best definition of theater costs is total project dollars divided by the number of seats.

Later, I was asking her facilities manager (whom she also subsequently fired) why she had picked an Architect who knew nothing about theaters and he related this story: After a meeting with the Architect, one of her people asked her "how that feels". "How what feels?" "Having his head up your ass that far?" Her reply is a classic: "Actually, I rather like it." Needless to say, that is NOT a good way to choose an Architect.

Here is another story. Another one of my repeat customers was a small college. (I think there were _five_ presidents involved in my work there.) Anyway, an architectural firm claiming to be a specialist in libraries got in touch with their librarian and the timing was right – they needed a new library. One of the firm's claims was that he was a professor in the Architectural Department of a university and he could save the owner money by having his students do the drawings. Yeah, right! One of the first things that I learned about libraries was that the stack areas do not want sunlight and one of the first things I noticed about the design were the skylights to let sunlight into the upper floor where – you guessed it – the stacks were to be located. The second thing I noticed was that the stairways and center atrium would not meet fire safety codes. Not even close. Then I found that the elevation drawings did not match the floor plans. And the Owner was already asking me to price the steel. Later I found that the firm had done <u>one</u> other project, a ***museum*** in South Bend. When I faced the principal of the firm with my criticisms, the librarian was offended and never forgave me. Later a new president built the library in the conventional manner using a qualified Architect that showed them a library they had done. They spent over *six* million for less building than we had offered for *four*.

But wait! Stop foaming at the mouth. ***Spending six instead of four is a good thing for a college president!*** He and the board of directors will be able to brag about their SIX million dollar facility! *

So, I have digressed. But I think that these stories say more than meets the eye. Like about my experience as well as my stubborn streak. And that I really *do* hate waste. Anyway, there are clues there on some of the ways **NOT** to choose an Architect.

*See "basements".

Usually, the way you DO pick an Architect is to ask several to make a proposal. This is what we used to call a "Dog and Pony Show". Sometimes they even serve refreshments. The most successful will be the one that has the best "Pitch". This is not all bad, but on the other hand, it isn't worth much to you and your project (unless you are in the advertising business and simply want to learn more about "pitches"). Of course, in the first place, somebody had to make a list of the firms that would be invited to make presentations. I'll just skip that little problem for now.

But DO try to include at least one local firm. No matter how small they may be. A firm that your friends know. A firm that knows the area contractors.

Those interviews are usually followed by visits to see their projects. Of course those will be only the projects that they want you to see. And if you talk to the Owner of that project remember that he usually is not eager to admit a mistake. If you **do** see a project that you like, be sure that you are going to get the same project manager within the firm.

I remember a manufacturer in Albion, Michigan that chose one of the most renowned firms in Detroit to be their consultant. So the relatively small manufacturer in little Albion got the newest member of the big consulting firm. That made the one-man local firm that already knew their business and their facility – the one they had discarded – look very good.

Now I'm going to make it worse. My stories and my advice seem to say that you should choose a Consultant with experience in your type of project. That is not necessarily true. Some of my most successful projects were the first of a kind. They included a college athletic facility, a meat processing plant, a large church sanctuary, a natatorium, and a racquetball facility. All it takes is some research. Besides, that is where you, the Owner, come in. You want to tell the Consultant what <u>you</u> need in <u>your</u> facility. The knowledge they have gained from other kinds of projects will be useful in yours.

There are other things that are more important. Like knowing construction costs.

Finally, I have used that word. "Construction". Which leads me to the word "Contractor". Capitalized. There are two misconceptions in the business. The word "Architect" is not spelled G-O-D and "Contractor" is not spelled C-R-O-O-K! I've already discussed the first. Now the second: Sure they are businessmen, but that doesn't make them crooks. *You* are probably a businessman. The big Contractors have too much to lose and the small ones don't have enough to gain.

Do you know where this myth comes from? It is from the adverse relationship that is established when the Architect has to enforce his standards on the Contractor who has to be the low bidder to get the job. The Contractor is the fall guy and is suspect because of what one of my good friends used to call the "Bid, Chisel, and Pray" system. You "Bid" the job, "Chisel the subs, and "Pray" for dozens of things like good weather and no strikes. The Contractor is the one that puts his money on the line on every bid. Remember this when we discuss other types of contracts.

But we haven't finished "How to Choose an Architect". Why don't you ask the Contractors who the best Architect would be? I've never heard of anyone doing this unless he or she is already a client of a contractor. I know, you don't trust Contractors. Well, you don't have to. But you can. They won't lie about the Architects they have worked with. Not unless they fear that you are going to tell Architect "A" that Contractor

"C" advised you against him. (Because the Architect can cost the Contractor a lot of money in this very flawed system.) By asking the Contractors, you will learn which Architects you can trust and which ones know what they are doing and something else that most people don't realize: which Architects respect Contractors and work with them instead of against them. That is so important that I'm going to repeat it right now. Choose an Architect who will work *with* the Contractor. Because your *Contractor* will become the most important member of your building team.

Okay, now that you are beginning to think about a CONTRACTOR, here is my best advice: CONTACT A CONTRACTOR FIRST! Before you waste any time trying to figure out how to choose an Architect.

Let's lay that thought aside while we discuss the second method for getting a building built.

An Alternative

In recent years most small school systems have been using a system called "Construction Manager".

By definition, somebody, usually a General Contracting firm, is selected to be the Owner's representative and manage all phases of the construction. That *could* be really good because a firm that knows construction better than anyone would help choose the Architect, determine the budget, supervise the design, prepare the estimate, make cost reduction suggestions, take bids from sub-contractors, and supervise the work and the payments.

They work for a percentage fee, sometimes capped, or a fixed fee (which, of course is based on the thought of a percentage). The Owner can keep as much control as he wants. Small school boards are suitable clients for this method because they need a "Professional" to turn to when they get public criticism.

The downside is that, usually, the Construction Manager does not guarantee the price. And the Owner has to make monthly payments directly to every sub-contractor. The Owner is, in effect, his own Contractor and has all the risks.

But, as currently being practiced in our area by the school districts, this method is a **HOAX!** First, the Owners often pick the Architect without the Manager's advice. And the Construction Manager is picked the same way as the Architect – the "Dog and Pony Show". Both selections are often swayed by a low percentage fee. (Once again, that is a percentage of the cost, so it is in direct conflict to saving money.) In the Construction Manager's case, that fee can get very low simply by charging everything as "costs" to which the fee is then applied. They then argue that a General Contractor would ask a much higher fee when, in reality, *that* fee would include a lot of overhead costs. Then, there is almost no effective plan review. The Architect and the Construction Manager become partners and studying the plans to make improvements is WORK that could conceivably reduce the cost and hence the fee. All oversight is lost. And *often*, amazingly, it is the Architect who provides the estimates.

Even worse is the way in which the budget is established. Since this part is being done *before* the bond issue is passed, the Consultants are working on speculation. They do not even know *when* a bond issue *might* be passed. They can afford little time to get accurate estimates. So they work on historic "rules of thumb" that often have little resemblance to the particular project. I have seen some of these that were atrocious! Their main concern is to get the budget high enough that it will be easy to stay within. Their only risk is that the bond issue won't pass and they will have *no* job. But that risk is worth taking so that they will not have to WORK within a tight budget. Besides, the Consultants have a contract for the **project** and the school will still need the work done and they will keep trying until they get a bond issue to pass. Then the CM has a gravy job. All he has to do is stay within his own extravagant budget.

Then, usually, they put a young college graduate engineer on the job as a coordinator instead of an experienced "superintendent" who will "kick ass".

So the sub-contractors, who have now all become "prime contractors" (and, incidentally, have to be individually bonded), mark these jobs up higher to make up for the inferior coordination that they have learned to expect, whereas if they are working for a General Contractor, they know that he has a financial interest in getting the job done in the shortest time.

There it is again – the word "CONTRACTOR" as in "General Contractor" because they are not to be confused with Defense Contractors or Labor Contractors. We should say "Building Contractors" or "Builder" but you know the subject of this book so I'm going to shorten it to "Contractor". And I'm going to capitalize it! Of course I'm prejudiced. But every word I write is true.

Next we are going to the third method for getting a building built. But first I need to point out an enigma about Construction Managers. After I watched a particular firm screw our local district as described above, that *same firm* was hired by a large commercial company to be the Construction Manager for a multi-story office building. Private money. The project "came in under budget and on time"!

How could that be? Well, this owner had their own construction department and they knew what to demand of the CM, including the qualifications of the personnel to be on their project. They knew when the building would be built. (And they probably did more of what a construction manager is supposed to do than the CM did.)

[As usual, we should ask: who determined the budget and the schedule? Because, as usual, if you get the budget high enough and the schedule long enough before you start, "on time and under budget" will be easy.]

Another Alternative

So here is the third (and, I think, the best) method to get your building built. It is called "Design-and-Build". CONTACT A CONTRACTOR FIRST – and you won't have to contact anyone else! This is called "single source responsibility" and it is impossible to "pass the buck". But be sure it is a "*Design-and-Build* Contractor" (whatever I said above notwithstanding). You will have a real "General Contractor" who will be a real "Construction Manager" and will save you money with a practical design -- AND will guarantee the TOTAL price!

Incidentally, the *best* "Design-and-Build Contractors" are also "General Contractors" in the sense that they are staying in the conventional bidding process. This keeps them in touch with new ideas from lots of Architects as well as Sub-Contractors that they might not have known about. So a *good* D&B Contractor will be a General Contractor. A *real good* D&B Contractor will have the forces to do some of the work himself. But an *exceptionally good* D&B CONTRACTOR will also be competitive in the normal system.

At one time there were not many qualified D&B Contractors. My former boss initiated the concept in this area over 40 years ago and we became adept at the most sophisticated projects. He started with small manufacturing facilities by applying his knowledge of construction costs and common sense to save the Owner money. When he was the low bidder on a local church and they had the perennial problem of the bids coming in way over the Architect's budget, he advised them to dismiss the Architect. He then applied his common sense to their needs and built the same size building. It had plenty of aesthetic value and was within their budget. That was the first of many.

Now, even federal government jobs are sometimes handled as "Design and Build" and there are a lot of Contractors who advertise "Design & Build". Some of them even have those words in their name. But don't let the name fool you. If your project is not too sophisticated you can find several and they will even bid competitively. Even if there is only one who can handle your project, do not let the lack of competitive bids deter you. You will still get the most for your money, BY FAR, using this method.

Here is how it works: **You CONTACT A CONTRACTOR (a Design-and-Build Contractor) FIRST.** (Okay, I promise not to say that again.) Anyway, all you need to tell the Contractor is the performance that you want your building to accomplish. The less you tell him about *how* to do this, the more you can gain from his expertise. And the sooner you go to him, the better off you will be. He can even help you find a site.

The very BEST part is that, often within a week, he will give you a *firm, fixed, **guaranteed*** price AND a completion time for a facility that will satisfy your requirements. His price will include all design and consulting fees and all permits. There should be no extras later unless you change your requirements or insist on more expensive ways to satisfy them. It will be what is called a "turn-key" price. When he completes the project, and maybe sooner if it will help you, he gives you the key, you turn it in the lock and start using your building. He will probably be able to advise on utility fees and other unknowns. He will be able to tell you how he plans to satisfy your needs. He may even have preliminary drawings for your approval. And, for all this, you will not owe him a single Dollar!

All he asks is that you do not peddle his solutions to his competitors. That is very unethical. (That is why he may be reluctant to give you drawings at this stage.) He will be happy to discuss how he plans to satisfy your needs and if you don't like something, he will change it. You will negotiate a final contract. Maybe you can negotiate an "incentive contract" where you both share in the savings. That way he will not be afraid that you are going to make unreasonable demands.

He will then choose the consultants he needs for code-required registrations as well as expertise in design areas that he is weak. I have had as many as six registered professional consultants on a single project. Since I was directing the design and providing the details as needed, and since I needed no specifications or other bidding documents, the total cost of all these consultants was never more than 2% of the cost of the building. Usually it is closer to 1%. And that is included in the Contractor's bid.

Did you make note of that? 2%! As opposed to 5 to 7% for the typical Architect. 2% *only* if you sign a contract for a completed building as opposed to 2% for preliminary drawings for a building of undetermined cost. And that 2% is *not* added to the cost of the building – *it is already included!*

For a very sophisticated or special purpose project the Contractor may have already consulted with one or more specialists before making his proposal. This, too, is on a speculative basis, costing the Owner nothing if that Contractor does not get the contract.

Since the better D&B Contractor is knowledgeable about the costs of all trades, he will not have needed to take bids from Sub-Contractors and will not be obligated to any "low bidders" when he lets the contracts. He will be able to negotiate the sub-contracts at the appropriate time. He will be able to take advantage of their ideas in the areas of their expertise. They, in turn, will be able to do their portion of the work efficiently and expeditiously. That is why we don't need "Specifications" in which somebody with no experience in the field tells the tradesman how to do his job. The only "Specifications" used in this system are "Performance" where the Contractor has defined the Owner's performance requirements. That is also why *you* don't need competitive bids – the Contractor is getting competitive bids for you.

It gets better! This type of contract is most suitable for "fast tracking". Because the Contractor knows how the entire facility will be designed, work can often be started before the drawings are finished. Site work can be bid and started almost immediately. Often a "footing permit" is issued without complete drawings (at the "Contractor's risk"). I often ordered the steel joists and roof deck within a week after signing a contract. (Because they were in the "Critical Path" of the schedule.) And since the budget is already established for every trade, you don't need to wait for bids for the latter stages of the project. I never bothered to take painting bids until the building was ready to be painted. Then I only got painters who were ready and eager to do the job now!

The Contractor is willing to take these risks because time is money for him as well as for you. Especially if he is trying to beat the weather. He wants to ensure that you will be another repeat customer.

Compare this approach with the conventional system. After you select an Architect, he puts it in his schedule. (Worse yet, he pulls someone off another project to *start* yours – and slows down *two* projects.) When it comes up on his schedule, he prepares preliminary drawings, meets with you, probably goes "back to the drawing board" to revise them, and then – sometime, finally gets you to "sign off" on the preliminary design. He then spends months preparing the working drawings and specifications which are needed to control the quality of the work that will be done by the "lowest bidder". Then he advertises for bids and a normal bid time is about three weeks.

The three weeks bid time alone is more than the Design-and-Build Contractor needs to design *and* bid most projects.

Here is another irony of the conventional system: For a fee of 5 to 8% of the Contractor's price, the Architect will spend months drawing dozens (or more) of "blueprints" and he will print hundreds of pages of specifications that are "boiler-plate" in his office — not even specific to the project. Then he gives the Contractors **three** weeks to study those drawings and specifications so they can figure a *competitive* price that will allow them to comply with every word of those specifications and every detail in those drawings. (Words and details that they haven't even had time to study.)

The Architect will be the "judge and jury" in all matters of interpretation of those bidding documents. He will also make the decisions as to whether the quality of the work is up to "industry standards" and he has total control of the payments.

But he has **_no_** money at stake!

Then the Contractors will bet their whole business that they are right and that nothing will go wrong such as one of their major sub-contractors going belly up.

There is a little known side effect with a D&B Contractor. He can afford to spend a lot more time on the details of your project than the Architect would. Because when he designs cost savings he is rewarded by sharing in those savings, rather that being penalized by losing a percentage of those savings. So he has a bigger motivation to work much harder on the design. And DESIGN is where it's at!

So, is there a "downside"? Only this: You have to trust your Contractor. And I know that is not as easy as I wish it were. He may not trust you, so why should you trust him? Because you have to. It is worth it to get all these advantages. You can make it easier by simply looking at his other projects and talking to the Owners. True, you run some of the same risks that you do in selecting an Architect. But not quite. Owners are not nearly as reluctant to bad-mouth a Contractor as they are an Architect. You don't really need to be looking at his other designs because this design will be based on your needs and wishes. All you want to know is "Can I trust him?"

But here is what I think is the most important question to ask: How many repeat customers does he have? Or, if the customer only built one facility, would he go back to the same Contractor for his next one? I think that is all you need to know.

Here is another idea: Ask the sub-contractors who have worked for him. They know him best.

There is one other *fear* of a "downside". This *seems* like it would be very important. If the Contractor is guaranteeing the price and controlling the design and supervising the work, where is the quality control? Where is the oversight that the Architect promises? Isn't the fox guarding the henhouse? Good question!

My answer is not going to sound very good. You have the right to control the quality. But you probably do not have the time or knowledge. The building codes prescribe a lot of quality but the inspectors are seldom on the project.

The design consultants are your best assurance because their licenses and reputations are at stake and they must follow the codes (which, nowadays, cover just about everything). But they don't spend much time on the job either.

<u>Surprisingly</u>, in my experience, it was the <u>LEAST</u> problem we ever had. The reason is the Contractors' reputation. They have to be concerned about their liabilities, their guarantees, your repeat business, your recommendations to future clients — as well as the other things mentioned above. So there is my answer: In the course of 40 years and hundreds of projects, it was **never once a problem!** Just choose your contractor as carefully as you would choose your new car.

Time for a Summary

The secret of reducing Cost? – "DESIGN is where it's at!" And who knows construction costs best? The builder. So who should design your building? Simple. The builder. A DESIGN-and-BUILD CONTRACTOR.

"Time is Money." And what type of contract gets the building completed the fastest? Design-and-Build! And what type of contract has the least "soft cost"? Design-and-Build!

If you are still not convinced, or if you can't convince your board, (that *is* often a problem) you can still use an Architect instead of a D&B Contractor. But at least try to find one that will *claim* that he doesn't like to throw money away. Look for some incentive based contract. Then don't let him write "closed" specs that give one supplier a chance to inflate his price. And don't ask for complicated alternates that are made up of several sub-contractors bids or take bids in the afternoon or otherwise treat the Contractor like shit.

Notice how I slipped in "take bids in the afternoon"? We always said that every Architect should visit a Contractor's office at "bid time" so that he would see first hand the problems that he has helped create.

Let me try to describe it by using a hypothetical three million dollar job.

But first, let me try to describe the extremely high risk that is involved. More often than not, the Contractor is about to submit a bid that is several times more than the total net worth of his company. A mistake in arithmetic or a mistake in judgment or an overly optimistic guess of how long the job will take – or the failure of one of the sub-contractors whose bid he uses – or bad luck with the weather – or an accident on the job – or a strike – or an unreasonable Architect – or an Owner who doesn't pay – or any of several other problems – can easily *cost him his entire company!* But there are only so many jobs to choose from and he has invested a lot of time and money in the investigations and estimates required for his bid – so he *has* to be **low** bidder. Second doesn't pay a dime. On public work, he will have to submit a "bid bond" guaranteeing that is he backs out of his bid he will forfeit 5% of his bid. That alone, could force him into bankruptcy.

So, anyway, it is 1:30 and the bids are due at 2:00. The bids are to be delivered to an office 40 miles away so we have sent a "runner" and he has found a pay phone and has called in on an unlisted number. (The "runner's" job is so important that our Comptroller usually did it.) He is "sitting" on the phone because other bidder's "runners" are looking for a phone. We're lucky today because the runner is only 6 minutes from the desk where the bids are to be received.

*[This was written before cell phones were common, but they are not the perfect answer. The rest of the scenario is still a problem – actually, **many** problems.]*

The FAX machine is running steadily and all three of our other lines are busy with sub-contractors calling in their bids at the last minute so that it is too late for anybody to "peddle" their bids to a favored sub. These calls are being taken by people that know enough to ask if the bid includes labor and sales tax and then get the caller off the phone so that somebody else can get through. But they do not know the job well enough to ask critical questions that are unique to every job. Like if the bidder is using the "closed specification" product. (Mandated by the Architect.) The bid takers are handing the bids to the chief estimator's helper. The helper is trying to give the estimator only the bids that appear to be lower than the ones he already has. The estimator is trying to stay cool – very cool – and he asks only the most pertinent questions.

Of the 35 categories of sub-contractor bids we are using, there are as many as 10 figures (the biggest ones) that are varying widely so that we have **no idea** what our final cost will be. We are running out of time so the estimator finalizes the alternate quotes (mandated by the Architect) based on the alternate bids from the latest apparent low bidders. He now picks up the phone and punches the line that the runner has been holding and starts giving him the information that he needs to fill in the smaller blanks (often on two copies that must match). That will include the names of the major sub-contractors that the Contractor agrees to use – the ones that are apparently low. (This requirement, mandated by the Architect, is intended to keep the low bidder from "shopping" the subs after the bids are closed. Watch what happens.) Now there is only one figure to finalize. We have a running total cost. We can't decide how much to mark it up until we see the *final* cost because we know the competitor will use a percentage of his estimated cost – probably less than 5%. (Less than his overhead last year.) It is now 11 minutes till 2:00.

And here comes another bid! Electrical, the second highest sub figure in our total. And it is $38,943 lower than the figure we are using. We have never heard of this firm, we don't even know if he is Union, (we are a union shop) but he is from the same area as one of the competitors and that is why we don't get his bid until late. We *have* to use it! Especially since it is probably still higher than the one he gave his friend. So, we cut $40,000 from the figure we just decided on. The estimator picks up the phone, rounds the number off to something easy to write out in words (mandated by the Architect) and gives it to the runner.

He stands up, takes a deep breath, stretches, goes to the kitchen, pours a cup of cold coffee. The runner will call as soon as the bids are read. There is nothing more to do.

After 5 minutes, he ventures to take another look at the late bid. The low bidder's alternate bid is $5,000 higher than the bid he used to calculate his alternate bid so if the Owner takes the alternate, he's out $5,000 right out of the gate. And, of course, there wasn't time to change the listed name of the electrical contractor he promised to use. So, if we are the low bidder, we will be **<u>forced</u> to "shop"**, hoping our listed sub will reduce his figure to meet the low one. Another of the many ironies in this business.

At 2:21 the runner is on line three. Reads the low bidder's name and number. We are second. Pretty good, out of eight. (Actually second is the *best* bid. If you are *low*, you probably made a mistake.) But second won't pay the rent.

Anyway, I guess we won't have to worry about that electrical bid – or that alternate. Just put it all in a file, take the rest of the afternoon off, start on another one tomorrow.

When I look back, I really do not know how we did it!

It just seems like they *want* the Contractor to make a mistake. But suppose the bids were taken at 7:00 like most schools do. The phones stop ringing at 5:00 because the sub-contractor's estimators want to go home. There is time to check your figures. There is time to talk to the Electrical bidder.

You even have time to go to the bid opening yourself and experience that thrill. (An opportunity not relished by all estimators. I liked it. Especially when a bidder hears them read an exceptionally high bid and panics and runs up to the front to snatch his bid before it can be read. I've seen that happen. And, when collared out in the parking lot later, he admits that his figure wasn't even the lowest. Embarrassing? Not when you think about what I've been saying. Another time I enjoyed was when I spread the word that I would not be bidding and then showed up at the last minute to steal the job from the Contractor who thought he was the only bidder.) Then you might be able to go have a drink with one of your competitors and see which of you can tell the biggest whopper.

And suppose the Architects just asked for one figure? It could be written out in words as well as numbers like a check, but only one copy has to match itself.

And suppose they got out of the "no sub-contractor shopping enforcement" business? If it makes any difference who the major sub-contractors are that the general plans to use, they can ask the bidders before awarding the contract. That would give the Architect a chance to ask questions, voice his concerns, and negotiate a change.

Let's talk about those alternates. Because I think it is usually the Owner that requests them. (If the Architect had any balls, he would advise against them.) In their simplest form they are used to determine the comparative cost of one item or brand over another. They can keep "closed spec" bidders honest. Those are not too difficult to handle.

But, much too often, they are used to "balance the budget". A room or a whole wing of a building can be added or elimi-nated and, with every bidder submitting a figure, there is no danger of the contractor inflating his "add on" or deflating his "deduct" after he has seen the other base bids.

The problem is that a chunk of building requires another whole take-off and complete estimate and involves almost every sub-contractor. So one sub-contractor may be low on the base bid, but not low if his alternate is taken. I tried to use net figures to establish my real alternate costs, but my fellow estimators laughed at me. It really is not worth it. And, ironically, if alternates *are* taken, it provides an open door to "shopping".

My contention is that if those extra rooms are really needed, they should be in the base bid. If they are not needed, *they are not needed*. That's it! See why I said "If the Architect had any balls, he would advise against them?

If you insist on alternates, you should know this. You will probably get most of your money's worth with "Add" alter-nates. You will not with "Deduct" alternates. For one thing, the mark-up and the bond cost will simply be ignored – in other words left in the contract rather than given back.

Part II – What You Should Know

So what should you, the Owner, know about construction design? As much as possible!

Size does count! Let's talk briefly about the size of your building. One of the ways an Architect establishes the square feet required is to list the rooms, determine how many square feet each room requires and then add rule of thumb factors for walls (thickness) and circulation (corridors). That may be okay to *budget* the square feet, but you can see that the final design should have fewer, thinner walls and circulation plans that reduce the amount of corridors. One of my favorite principles is to always "double load" corridors. In other words, make corridors serve rooms on both sides.

Fewer, bigger rooms. Every room will have wasted space in the corners and, sometimes, against the walls. Every additional room means less flexibility of use and more corridors to access them. So, the less the building is divided into rooms, the smaller the building can be.

Fewer, better located doors. Rooms will have doors that usually require a storage area (when swung open). If the door swings into a corner of a room there is no storage area used. If there are two doors located diagonally opposite in a small room, the traffic pattern can decimate the use of the room. So, judicious location of the doors can make a smaller room just as functional as a larger one.

12 inches equals one foot. Sometimes there are ways to shave dimensions. Library stack dimensions, for example, require 57 inches center to center of book stacks. So, since that's almost 5 feet, almost all library designers will make it easier (on themselves) and use a 60 inch module. That negates a <u>potential</u> savings of 5% of the floor space. But be **very careful** with this type of thinking. The tight dimensions will not allow for columns between stacks or for any part of a wall thickness. And there will be five stack aisles in 23′ 9″ which will complicate modular design and such things as lighting layout. Also the module chosen will be used in the rest of the building which (even in a library) is a lot more than just stacks. In this case it is **far more important** to be sure that the building will accommodate a multiple of **full modules** so that all stack aisles will have books on both sides ("double loaded").

But it is easy to cut 2 or 3 inches out of every 9 or 12 feet. In offices, this will never be missed and has the added advantage of guaranteeing that you will never have a ceiling or floor covering dimension that is 12′ 1″ which requires the purchase of at least 14′, maybe 15 or 16′ of floor and ceiling material.

Common sense is your best tool in deciding the size of your building. And it is important not to scrimp in this area. Remember that a lot of the permanent value of a building is based on its size. And the **real cost** (as opposed to the estimated cost) of **added** square feet is probably half of the average square foot cost.

Plus – a larger building can give you room to grow and, hopefully, more flexibility.

So, enough about Size. SHAPE is much more important.

Unlike women, building materials are not intended to be curved! There are **no** building materials that are meant to be curved. Not wood. Not steel. Not glass. Not even concrete because it has to be placed in forms and it has steel bars inside it. Nor plaster, because it has to be supported by something which is cheapest when made (and shipped) straight. The only curved building component that I can think of that is economical is a round column. So, if you want to think curves, start thinking double the cost and then double that! Because you will pay dearly.

You are saying **"Lots of buildings have curves."** Yes, they do. And those buildings cost a lot more money. I'm here to help you avoid as much cost as I can. You may also be saying "What about arches?" Yes, arches were common for thousand of years. **When LABOR was cheap!** Before steel lintels or curtain walls.

There are also side effects. The most common curve in construction (and the cheapest) is a horizontally curved exterior wall. But that means that the edge of the roof is also curved. So the roof frame is either curved or in short angled segments. And the structural members perpendicular to the wall must vary in length.

And the roof decking and the roof insulation and the roofing material, all of which come in rectangular sheets or rolls, must be cut on a curve (field labor) and the outer portion that is cut off is discarded (a nice word for **wasted**). **Also**, the inside of the room will no longer have a nice straight wall to use – you can't even run pipes along the wall.

If you MUST have curves, they should be as **big** a radius as you can use. Then they should be made of the **smallest** standard straight unit you can use such as brick. And hope that you don't need to curve any steel members to support those wall elements.

Art class is over. GEOMETRY class is now in session.

As much as half of a building's total cost is in the "Envelope" which consists of the floor, the roof, and the perimeter walls. The floor is pretty much a fixed cost and is based almost solely on square feet. Some do not even consider it as part of the envelope. The roof has a few more variables that we will deal with later, but still, unless we are considering a multi-story building, the cost of the roof is based on the square foot size of the building.

The perimeter walls are an entirely different matter. And they are by far the most expensive component **per square foot.**

Okay, here we go: Square feet (of walls) = length x height. The height is determined by the interior usage requirement plus the thickness of the roof system. It may be one or two **dozen** feet high. But the length will be **hundreds** of feet. The length of the walls will be determined by the area to be enclosed and the **SHAPE** of the enclosure.

A circle is the **shape** with the most area for a given perime-ter. But, we know that a circular building is not cost efficient. Next best would be a polygon but that would necessarily be made up of triangles which would be inefficient roof framing. Economy in construction **begs for rectilinear**.

The most efficient building, therefore is a **SQUARE**. A build-ing 100' x 100' encloses 10,000 square feet using 400 linear feet of wall. A building 50' x 200' encloses 10,000 square feet but requires **500** lineal feet of perimeter walls (and foundations). If we apply some arbitrary numbers (exagge-rated on the low side) we have added at least $1.00 per square foot to your building cost.

Of course, there are other considerations. The first is that you may simply **want** a long narrow building. I had a customer many years ago who wanted to build his dream building for a screw machine shop. He envisioned his screw machines all lined up along one wall facing an aisle and on the other side of the aisle would be a storage area for the raw bar stock. The storage bay would be served by a traveling bridge crane. So he wanted a building 72' wide and 240' long. Since the storage bay would not be filled for the length as determined by the machines, I suggested that he build a building half as long and have two rows of machines against the outer walls with the crane/storage bay serving both rows of machines from the center. My design would then require a building 120' wide and 120' long. Not only would he save 2,400 square feet of roof and floor and 144 lineal feet of perimeter wall, he would save 120' of very expensive crane runway structure, rails and wiring. And I argued that he would not have to walk as far from his office to the farthest machine. I argued until I almost lost the job. We built it **his** way. It was his prerogative. (And then we built all four of his future buildings **his** way.)

*[You may have noticed that the hardest thing for me to accept as a design-and-build contractor was to give the customer what he (or she) wanted, even when I **knew** they were wrong.]*

Another consideration is the type of construction you choose. Thinking in terms of manufacturing facilities, there are two choices: a sloped roof (usually metal) with all drainage to the side walls or a "flat" roof (usually membrane) with an interior drainage system. With a sloped roof, you should not build an addition along the eave side because that will create a valley that will guarantee leaks. So, if other factors suggest a building 100' x 200' and if there is even the **remotest possibility** that you **or** your successor will

ever build an addition (in other words if the site will allow it) consider a building that is 200' **wide** (along the rake side) and 100' **long** (along the eave side). That will enable you to double the building to 200' square without a valley. Note that the "end" walls are higher at the peak so you have added at least 400 s.f. to the initial wall cost. In this case I would recommend a compromise to 160' x 125' for several reasons, **only one of which** is that the original perimeter is closer to square saving 5% of the initial wall cost. But if you have chosen a "flat" roof design, **anybody** will be able to add **any size** to **any side** at any time. [More on "flat" roofs later. Please do not decide yet.]

Still another factor is the "bay" size or module or column spacing that you **need**. More on that subject later, too, but at this time suffice it to say that if you decide to use 30' bays, a building dimension of 100' is not going to be a structurally efficient design. (Need I mention that uniform modules are important?) I once had a customer who wanted to build a speculative "shell" building to entice manufacturers to their industrial park. They asked for a 100' x 200' building. Since the most efficient structural dimensions at that time were 30' x 40' bay sizes, I offered a building size of 120' x 180'. Notice that both sizes have the same amount of perimeter. The efficiency of the structural modules made up for the added 1,600 square feet of roof and since there was no floor in the bid, I won the bid by offering extra square feet. The new Owner soon doubled the size of the building along the 180' wall and then **doubled it again** along the 240' wall. And it was easy because we had used a "flat" roof design and spent just a little more money to make the foundations ready for just that possibility.

While we are on the subject of "bay sizes" let's talk about "clear spans". Most factory Owners want as big an open area as they can so that there are few columns to interfere with flexibility. Indeed, this *is*, usually, quite important. But the longer the "span" of the structural members, the more the system will cost. On the other hand, almost always, there is little point in having the columns spaced far apart in **both directions**. (We used to joke that you needed a column every so often to hang the fire extinguishers on.) So be sure to analyze your needs and communicate them to your designer.

The good designer will try to use every structural component to its optimum efficiency. The most economical roof deck will usually span 6' (by using 3-span lengths). The most economical roof joists (so-called "bar joists") are 40' long. And the most economical beams to support the joists will be supported by columns spaced at 30'. The beams can be designed as "overhanging support" (commonly known as "cantilevered"). Then the steel can be purchased in 30' or 40' lengths which are stocked in all warehouses. Any deviation from these lengths will result in "drops" (scrap) or going to the next length. [Warehouses also stock 50' lengths and if there is time and enough repetition, steel of ANY length can be ordered from the mill.]

Thus the most economical bay size is 30' x 40'. If longer spans are needed, the cheapest components to change are the joists and it is not unreasonable to go to 60'. Longer than that, the trucker will have to get special permits to haul them to the job, probably through more than one state.

The most economical beams used in 30' bay spacings are usually 14" and there is a wide variety of 14" beams available. Staying with the same (nominal) depth of beams makes connections and other details easier. Less than 30' column spacing usually will not save money because columns will have to be added and their cost, coupled with the required foundations, will probably out-weigh the cost of the heavier beam that is required. There may be an exception if there is a traveling bridge crane that has to carry an extremely heavy load.

One of the more important considerations is the size and shape of the site. Most industrial sites will probably be large enough that they will not restrict the **shape** of the **first** building. But many commercial sites, because of land value, will. We had one case where the required parking area limited the footprint of the building and mandated a multi-story building. The Architect's solution was a five-story structure. **However**, I could have tweaked the dimensions a **little**, still had "double-loaded" parking aisles, made the dimensions of the building closer to square, and built the same **usable** space in four stories. Besides saving almost 3000 square feet of the highest and, therefore, the most expensive walls, there would have been untold savings in stairs, circulation, elevator stops, vertical chases, mechanical and electrical, even time. They would not have needed as much floor space because dividing the space by four is far easier than by five. There would have been lots more savings, too. (The least expensive ground floor would have been 1/4 of the total instead of 1/5.). But it was too late. The Owners liked the "Look" of the five-story building and did not have time to "send it back to the drawing board". Illustrating once again how important the original concept can be.

[The contract for that project was a "Guaranteed Maximum with a Shared Savings Incentive Clause" and I was allowed to re-design the structural system from "Simple Slabs" to "Composite Beam Design" resulting in a major savings and a *very* happy customer.]

An important decision in factory design is the height. Let me add one word to that statement: "CLEAR". The **only** definition of height that is meaningful is the "*clear* height". So pay no attention to the packaged building that describes the height as the "eave height". You need to know the clear inside working height for your operation. In some cases that may be a multiple of storage bins that you will be using. (And, in some cases, you may be able to accept the obstruction of a lower frame member at 20' intervals. But remember that you probably will have some piping that will not be able duck down and back up at each frame.)

Above 16', the cost of the height becomes very important. The column design has to be increased geometrically with the "unsupported" (or "unbraced") length. Also, the structural system that supports the walls is designed for the wind load and that load also increases geometrically.

On the other hand, if an Owner thinks he only needs a clear height of 12', I will still recommend 16' because of future unknown needs and marketability. And if part of the building can be lower, I may recommend raising it to the higher need (unless that is extreme, like a tower). Because the complications involved in a step will probably cost more.

Okay, geometry class is about over. Maybe.

Let's talk about **CUTTING CORNERS.** I don't mean skimping on quality – I mean design **with fewer corners. Because every corner costs BIG money!** Unfortunately, I cannot quantify how much more because I have never seen an estimating method that counts them and values them individually. They are included in the average historical unit prices of the wall. I think that if we *could* quantify the cost of a corner and add it as a separate figure, the cost of the rest of the wall could be cut in half. So, you **might** get a **bargain** if you have more corners than the "average" building, but **only if** you have let a contract based strictly on an estimate rather than the proven cost. Maybe not. though. The good estimator gives credit to a design that is **"clean"**.

Why does a corner cost big money? I suppose that is obvious, but I want to clarify it. **Because the builder has to STOP and change directions.** If you have ever watched masons laying block or brick, you may have noticed that one man is assigned to build a "lead". He will be one of the better craftsmen in the crew. He must make sure that the corner is plumb in both directions and that the height of the courses is being maintained and that the angle is maintained. Someone else has already taken a lot of time to "lay out" that corner. Then they string a line to the other corner and the other masons almost *throw* in the units "to the line". If that length of wall is not a multiple of 8" then someplace in the wall, they will cut (saw) and "discard" (remember, a nice word for waste) part of a piece.

That is only the outer layer of the cost of a corner. At every change of direction of the structural system there is also a cost. Extra corners invariably come in pairs, an inside and an outside and rarely do the **three** wall dimensions that are affected stay on the structural module that, hopefully, has been efficiently chosen. So two columns are added and neither one will be used to full capacity. (Because a column is designed for its height more than the load it carries.) And columns mean footings that are also able to carry more than they have to. And footings mean piers and anchor bolts and leveling pads. The only thing worse than a corner is a rounded corner which requires **two** columns **each** and some fancy framing and all of the other bad things about a curve.

Incidentally, I have been assuming that we were both thinking **square** corners. From what you have learned you can figure out for yourself the cost impact of an acute angle corner. Think about the end of a brick or block and pretend you are a bricklayer. **OUCH!**

I'm not suggesting that all buildings should be perfect squares with only four corners. (Well, I am, but I know better.) But you should realize that the **closer to square with the fewest corners** you can live with, the lower your costs will be.

The cost of windows is based on the same geometry. You get the most light per dollar with a square window because the perimeter of the window is a major part of its cost. You may have noticed that the windows in the building on the cover are square and yet I listed them as one of the offending costs. The reason is that there were so many and so closely spaced. The jambs in a masonry wall are similar to corners – start and stop. Of course, if a window is too wide, the lintel becomes a problem. (The best way to get natural light into a manufacturing building is a row of windows at the top of the wall so the roof edge framing will support the top of the windows.)

Time for another summary: **Over the years I consistently violated every principle I have espoused. But JUDICIOUSLY!** Because I knew the theories and used common sense and was willing to spend whatever time it took to get the most value for the money.

New subject: What is the main purpose of a building? To keep you **DRY** and **WARM**. And those are the two things that the builder has the most trouble with. Let's talk about **DRY** first. I already mentioned that there were two types of roofs: sloped and "flat".

Flat roofs are NEVER FLAT. Not really. Not anymore. When a roof appears to be flat it is actually sloping to interior drains that are connected to the storm drain system. The minimum slope that any roofing material manufacturer will guarantee is 1/8" per foot. That would be for a "built-up" roof which is the old "tried-and-true" alternating layers of felt and asphalt. Most "single membrane" manufacturers require 1/4" per foot. These slopes are accomplished by "warping" the structural design to create small peaks and valleys. Do not make the mistake of using "tapered insulation" to create slopes except maybe in small awkward corners. This is the easy way out of a lot of design work but is very costly.

Roofing felt is NOT waterproof. In fact, some are perforated or merely open fiber mats. The so-called felt is there to separate and reinforce the layers of asphalt or pitch which **is** the waterproofing.

"Gravel" is not ballast. The gravel on a "built-up" roof (which is specified to be placed at 4# per square foot) is there to protect the asphalt from the sun.

"Stone" IS ballast on a single membrane roof and is required at the rate of 10 pounds per square foot to hold the membrane down when there is not a mechanical fastening system.

[Notice that if you were to re-roof using a ballasted roof, your original building design might be 6 pounds per square foot under-designed and that *could* be enough to be fatal.]

"Rubber" roofs are not rubber. They are polymer or butyl or any one of a number of different compounds. There are so many that everyone in the business is confused. I know an Architect that went to a 3-day school to learn about single ply membrane roofs and he returned thoroughly discouraged because there was some drawback to every one.

Finally, **Roofs don't leak!** Well, not in the middle of the roof. They leak at the **holes** or at the curbs and flashings around the holes or at the walls between different roof heights. That is one reason to **not** have skylights. (The other reason is that they add a terrible load to the cooling system.) Anyway, the designer and the roofer must take the greatest care to make these places water-tight.

My best advice: First, I want you to know that I am out of date. When I left the business, there were more and more advances in roofing materials coming on the market. More roofers were switching from "built-up" to single ply, partly because the "built-up" required extremely high field quality control. And the whole idea of ballasted roofs has probably already been dropped as better mechanical fastening systems were developed. So, what I was doing for my buildings was to find a **roofing contractor with at least 20 years experience** who could be expected to be in business indefinitely. Then I let **him** decide what kind of roofing to use after assuring him that I would pay for quality and long life.

Stay WARM. And keep your cool.

The largest section of a set of specifications is for the "Mechanical Trades". That covers Plumbing, Heating, Ventilating, Air Conditioning and Fire Protection. In the normal System where the Architect is controlling the bidding, this usually falls under one Subcontractor, the Mechanical Contractor, who has at least two major sub subs, a Fire Protection (Sprinkler) contractor and a HVAC (Sheet Metal) contractor. He will also probably have an Insulation sub and a Control systems sub. The Mechanical Trades subcontractor is historically a plumbing contractor because most heating systems used to be hydronic, either steam or hot water. As ventilating and cooling and humidity control requirements have grown more important, the plumbing trade has become smaller than the HVAC. So one of the first things that a D&B Contractor does is to separate these trades and deal with each directly. This is for two reasons: They avoid the extra mark-up that the plumber adds for his subs. More importantly, they gain the control of choosing and dealing directly with the smaller contractor who is an expert in his own specialty.

Bear in mind that the Mechanical work will usually be about a third of the total cost. There is not a lot to say about the plumbing or the sprinkler design. Both are pretty straightforward and are closely controlled by codes. But the HVAC design has a lot of pitfalls and a lot of ways to waste money.

First, I should point out that the four words "heating, ventilating, and air conditioning" are redundant. When air is heated or cooled it is being conditioned. And when a building is ventilated, the air inside is being conditioned. When the air is filtered or humidified, it is being further **conditioned.**

Now, are you thinking **AIR**, moving air, circulating air, distributed air – ductwork? Because I want you to realize that if you have to do **anything** to condition the air that goes into every room, you might as well heat it, too. Then you can **skip the hot water heating system**. You can skip the pipe-fitter and the pipe insulator and the controls for that whole system. In a more sophisticated system, you may want to heat the air source with a hot water heat exchanger or maybe even put some hot water coils in branches of the ductwork for localized control. But, please, **BEG** the designer to **keep it simple.**

And that is the most important thing I can tell you about Heating and Cooling systems. **Keep it simple.** Especially the control systems. Fancy control systems will be your worst nightmare. Do not even attempt to give each room its own temperature. In my experience, it can't be done. The more "sophisticated" (a polite word for "complicated") the control system is, the more problems there will be. I guarantee it! All contractors will guarantee it.

There is so much more to say about HVAC systems that I don't know where I would stop. But a few thoughts won't hurt. Watch out for buzz words like "energy efficient". It's a good thing and is often mandated. But try to combine it with another buzz word "life costing". In other words, ask if the savings over the life of the building (or maybe that should be the life of your ownership) is as much as the cost of the equipment involved? A good example is a "geo-thermal" system that squeezes heat out of ground water with a "heat pump". It usually requires two deep wells – one to bring water up – another to replace it. And watch for "re-cyclable" and "recovery" and "green". You may just have to let someone else save the planet. But there is no need to feel guilty. If your building is designed efficiently, you are already saving natural resources.

One more comment on the HVAC subject. When I asked my Doctor how many aspirins I should take in case of a heart attack since I am already taking one a day, he didn't know. But his answer was "That's why you are taking one a day – so you **won't have** a heart attack." My comment is this: **"That's why you have a Design-and-Build Contractor – he'll know what's best for you!"**

Site Design. I want to touch on this subject even though it is like every other aspect of design – just common sense. I have seen so much money wasted on extravagant, or more often, just careless, design. The importance of this portion of the work can be brought home with the experience I had with the theatre we built that was designed by the new Architect. The city had to re-zone the property that the Owner was buying so the plans had to be submitted to a review board that included a (dare I say it?) woman who was more interested in the proposed landscaping than most code reviewers would be. So before we were brought into the project, the Owner had found a large Professional Site Engineering Firm and offered him a sizable fee to do the site design. Everything that could go wrong went wrong. The site was 80 miles from their office. They were busy and this project was smaller than most of their work. The theater owner's representative was the financial officer and his office was 300 miles from their office and he didn't have the time or the know-how to kick ass. To top it off, the engineering firm always "subbed out" the landscaping to a "Registered Landscape Architect" (and the city did require "Registered"). When I got on the job, ready to start, pending city approval, winter coming, the site drawings were "almost done" but the landscaping specialist hadn't even been selected. So I kicked ass and expedited drawings "even if they are wrong" in the interest of time. Maybe you can imagine how bad they were, but I doubt it. They were drawn simply to get me out of their hair. There was no, absolutely not any, time or thought given to efficient design. Some things didn't even meet code and that was the main reason they had been hired. But those were the drawings we took to the City and got approval so we could start construction. And once approved you better not be changing them.

But I did. By cajoling and sweet-talking and meeting with the right guy, and pointing out flaws in the approved design that we should fix, and maybe just by becoming a nuisance, I managed to re-design every aspect of the site and saved over $200,000 (out of $800,000). And I gained parking spaces which was critical because the Owner had bought a site that was almost too small for the required parking and the retention basin. (Incidentally, that site precluded any thoughts of planning for a future expansion.)

Let me list the items that go into the estimate for site-work on a job like that. Stripping topsoil, grading, excavation, backfill, removal of excess or unsuitable soil, purchase of porous soil. Curb cuts and street curbs and de-acceleration lanes. The holding pond including excavation, lining, seeding, protective fencing. The parking lot including storm drainage, sub-soil compaction, gravel, bituminous or concrete paving, curbs and gutters, island plantings, striping and lighting. Sidewalks. Signage. Perimeter landscaping including topsoil and seeding or sod and maintenance for one year. And this category also includes extending the utilities (sanitary sewer, water, electric, phone) from the property line to the building.

So you can see that it is worthwhile to locate the building and route the utilities so they are at the right elevation when they cross and as short as possible – because they cost anywhere from $10 to $40 per foot. (At that time.) And of course it is important to slope the pavement to the drains and still have as few drains and as small a pipe as possible. Here are some other caveats: Plan for future expansion. Try to balance the cut and fill so that dirt will not have to be purchased or removed from the site. Use the building "setbacks" required by code for parking. "Double load" all parking aisles. Use perpendicular parking wherever possible because diagonal parking takes more space. Have as few cross-aisles and access drives as possible. Provide overflow parking along main aisles.

Basements: I had thought I had finished this book when a former customer contacted me and was quite upset that the building he had donated money for was going to cost double what he had thought would be generous. He described the building as "just a box" and I had to agree that his budget should have been enough. When, later, I got a little more information, I found that the Owner (not the Donor) had thought they could get some "cheap space" by having a basement. And the architect (notice *not* capitalized) – [and, incidentally the same one who had done the $4,000,000 library project for way over $6,000,000 (actually closer to $8,000,000 I now learn!)] **had not straightened them out!!** I suppose because he didn't know any better.

The naivete of several people came to mind, the first being me, because I had forgotten to include this subject in the book. Naivete is a nice word for stupid, but the architect was the one who should have known better.

It is a common misconception that comes from houses. And even then, **basements are never free.**

In a public or commercial building, a basement will more than double the cost of a single story building. When you think about it, you probably already know why. Maybe the biggest reason is that the main floor structure becomes a "supported" system that costs even more than a roof structure. The second biggest expense is an elevator which is required by Handicapped Accessible codes, even in walk-out basements. Then the Fire Marshall is going to require two exits that must open to the outside without going through any of the main floor or using the elevator. If it is not a walk-out, that means two stairways. And two stairways and an elevator are going to use up space on **both** levels! So if you were thinking of dividing your required space by two to put half of it downstairs, then you have to add to the size of the building to compensate.

You may say, "But what about those cheap basement walls?" Those walls probably cost *more* than the brick wall above ground because they have to retain the earth and they have to keep out the water. And you may ask "Isn't there less heat loss through a below ground wall?" Yes, but most of the air conditioning (that's heating, too, if you have forgotten) requirements for anything but a residential building are based on the *volume* and required fresh air changes. And there are additional square feet lost to vertical shafts for the ducts.

Now, **maybe,** if the basement is to be used strictly for storage and not "finished", **some** of these costs can be reduced or eliminated. Still, I would recommend that you add to the size of the main floor, which can then be converted to other uses at a later time.

What amazed me most about this Owner's mistake was that, just before the library fiasco, they had built a small office building and split the space into a basement and first floor. Not only were the elevator and stair requirements a **huge** percentage of the cost, but they were also a large percentage of the floor space because the building was so small. Splitting of the space also ruined the efficiency of the working layout. Even though the site *was* cramped, there **was** room enough for the building to be on one level. And they would never have had the drainage problems they had.

There is another problem involved in this latest project. This building is rather sophisticated, especially in its mechanical and electrical requirements. The Owner has arbitrarily decided that the dedication will be at Homecoming so that is a requirement that the Contractor will have to meet. Regardless of the impracticality of their demand. It is now the first of May. I thought to myself, "Why are Owners so stupid?" but then thought, "That's why I'm writing this book". Anyway, I told the Donor, "You should talk to my former boss – **NOW! – yes, this *noon*, you cannot wait another day! The project will *have* to be fast-tracked. That is your <u>ONLY</u> hope of getting your building on time and in budget!"**

Finally I have run down. But before I leave you on your own, here is a very brief list of some things you should NOT scrimp on:

Hardware and door quality.

Insulation (although there is a point of no retrun).

Electrical capacity.

And low maintenance finishes.

Glossary

Here are some things that you might like to know when you are talking about construction. This may earn you a little more respect from the Contractor.

1. "Concrete" is made from "Aggregate", usually stone and sand, and "Cement". "Cement" is the active ingredient and is activated by water. The concrete "cures" as the water leaves. Better concrete is poured with a "lower slump" which means "stiffer" or less water. There is no such thing as a "Cement" driveway.

2. "Asphalt" is one of the ingredients in "Bituminous Concrete", along with "Aggregate". There is no such thing as an "Asphalt" driveway. Bituminous concrete paving costs less than cementitious concrete and is more flexible.

3. A "Column" is a vertical beam but it is never called a "beam". And it is never called a "post" unless it is on your porch or in your basement.

4. "Wall-bearing" really should be "Load Bearing Wall". It usually refers to a type of construction in which the roof structure is supported by the walls.

5. The "Frame" of a building is its "Skeleton". It is usually steel, concrete, or "Timber". A "Rigid Frame" refers to the joints which are **designed** as rigid but are not really "rigid". (Anything that is truly rigid must break.) Common usage of the term "Rigid Frame" refers to "Pre-engineered" building construction and has the advantage of more economical long spans.

6. A "Ceiling" is the top of a room. The top of a building is the "Roof".

7. A "Supported Slab" is a concrete slab that is structurally supported, or self supporting, above ground as opposed to a "Slab on Ground".

8. "Masonry" usually refers to "Unit Masonry" and covers brick, stone, and block work. Most blocks are "Lightweight" concrete using cement and expanded aggregate. "Cinder" blocks are no longer made.

9. Blocks can be made with molded faces. There are also "Split Face" and burnished block and they can be integrally colored. These are all ways to up-grade an otherwise cheap building material. They can be up-graded to the point that they may cost more than face brick.

10. "Joists" and "Purlins" are secondary members that are supported by "Beams" or "Trusses". Purlins and trusses are usually used for roofs while joists and beams are used for both roofs and supported floor systems.

11. "Girts" are secondary wall members that are supported by "Columns". Girts support "Sheathing" or siding and *can* brace columns in one direction.

12. "Precast" concrete is cast off-site and hauled to the job-site. "Prestressed" concrete is commonly precast but can be "poured in place". The reinforcing cables are either "Tensioned" (stretched) before the concrete is cast around them or "Post-tensioned" after the concrete is poured.

13. "Tilt-up" construction refers to pouring concrete walls on the ground and finishing them as a slab, then "Tilting" them up to be supported by the building framework. This saves the huge cost of wall forming.

The use of 13 items in the Glossary is deliberate. But, lest it be too subtle, I will explain. My intent is to further my credibility by showing that I am not superstitious.

And now, before that rash statement has had time to antagonize you, let me repeat one more time:

DESIGN IS WHERE IT'S AT!

I hope I have been of help. Good luck.

P.S. If you are already in trouble, having hired an architect whose ideas are more than you can afford, contact me at

cococucopub@gmail.com

I might be able to arrange to review your drawings and make suggestions. It usually does not take me long to find ways to do what you *NEED* at substantial savings. The key word there, of course, is *NEED!* That would require you to be honest, not only with me, but with yourself. *IF* I think that I can help, I will propose to work for a small percentage of the savings.

I am available most weeks of the year and I will try to reply within three days.

(not used)

www.ingramcontent.com/pod-product-compliance
Lightning Source LLC
Chambersburg PA
CBHW051246170526
45165CB00004B/1591